Space, Stars and Planets

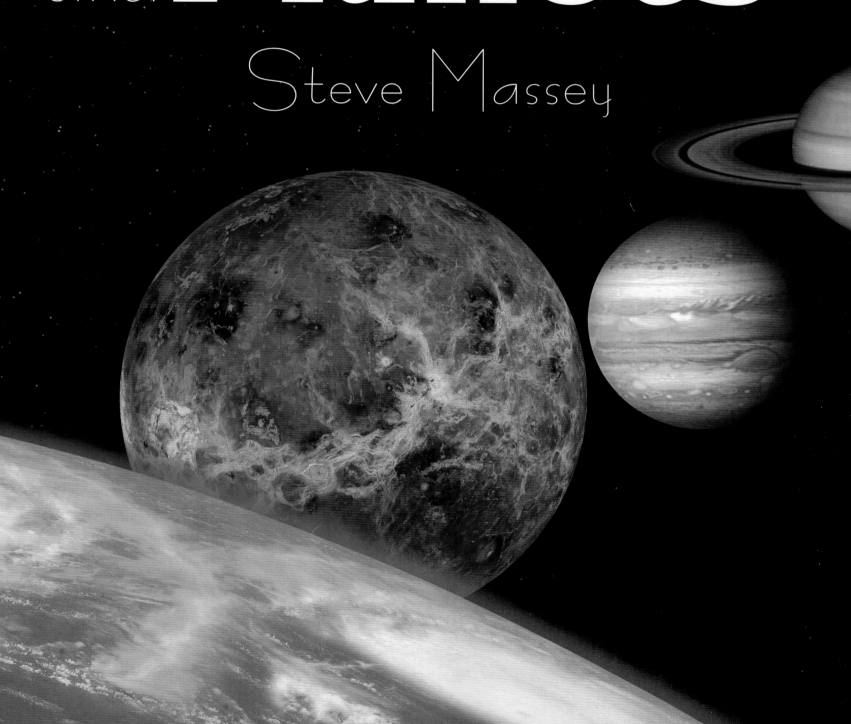

Space, Stars and Planets

Steve Massey

First published in Australia in 2007 by Young Reed
an imprint of New Holland Publishers (Australia) Pty Ltd
Sydney • Auckland • London • Cape Town

1/66 Gibbes St Chatswood 2067 Australia
218 Lake Road Northcote Auckland New Zealand
86 Edgware Road London W2 2EA United Kingdom
80 McKenzie Street Cape Town 8001 South Africa

Copyright © 2007 in text: Steve Massey
Copyright © 2007 in maps: Steve Massey
Copyright © 2007 in photographs: Steve Massey except where otherwise credited on this page
Copyright © 2007 New Holland Publishers (Australia) Pty Ltd

All rights reserved. No part of this publication may be reproduced, stored in a retrieval system or transmitted, in any form or by any means, electronic, mechanical, photocopying, recording or otherwise, without the prior written permission of the publishers and copyright holders.

10 9 8 7 6 5 4 3 2 1

National Library of Australia Cataloguing-in-Publication Data:
Massey, Steve.
　　Space, stars and planets.

　　Bibliography.
　　Includes index.
　　ISBN 9781921073069.

　　1. Stars—Juvenile literature. 2. Planets—Juvenile literature. 3. Astronomy - Juvenile literature. I. Title.

　　523.1

Publisher: Martin Ford
Junior Editor: Sally Hills
Designer: Tania Gomes
Production: Linda Bottari
Printer: Tien Wah Press, Malaysia

Picture Credits

Abbreviations: t = top; b = bottom; c = centre; l = left; r = right

NASA/HQ/GRIN: pp. 44–45
Harrison H. Schmitt/NASA/JSC: p. 24–25c
CSIRO: p. 35br
NASA/JPL: p. 7cr; p. 9br; p.15tl, cr; p. 19bl; pp. 20–21; p. 21bl (image compiled by the author); pp. 26–27; p. 27tr, bl; p. 28; p. 29cl; p. 32br; p. 33tl; p. 34; p. 38t; p. 41bl
NASA/ESA: p.16
NASA/GSFC: pp. 44–45
NASA: p. 1; p. 2–3b; p. 8tr; p. 9 cl; pp. 10–11; p. 11cl; p. 15bl; p. 17cl; p. 23; p. 36tl; p. 40br
NHIL: p. 7bl; p. 22
Steve Quirk: p. 12cr, bl; p. 18tl; p. 35tl; p. 37cr; p. 38bl, p. 39tr
Steve Massey: p. 2tl; p. 3; p. 6tl, cr, bl; pp. 6–7; p. 10 cr, cl, br; p. 11tr; p. 12–13; p. 16br; p. 17tl, cr; p. 18bl; p. 21tl; p. 24tl, bl, c; p. 26tl; p. 27cl; p. 29tr; p. 30tr; p. 31; p. 32cr, bl; p. 34bl; p. 36; p. 37bl; p. 38tl; pp. 38–39; p. 40tr; p. 42tl, bc; p. 43br
Steve Massey and Steve Quirk: pp. 42–43
SOHO/NASA: p. 17bl
NASA/STScI: p. 11br; p. 33
Robert Williams/NASA/STScI: p. 12
NASA, ESA, HEIG, and the Hubble Heritage Team: p. 8tr
NASA/Goddard Space Flight Center: pp. 10–11; p. 36
NASA/Kennedy Space Center: p. 9cl
Scott Massey: pp. 14–15
Steven Felmore and NHIL South Africa: p. 22cl
AVHRR/NDVI/Seawifs/MODIS/NCEP/DMSP: p. 23 cr; p. 35c
Harrison H. Schmitt/NASA/Johnson Space Center: p. 25
M McCaughrean, CR O'Dell and NASA: p. 7tl
Jim Bell (Cornell University) and STScI: p. 27cr
STScI: p. 33
Steven Voss: p. 28; p. 43cl
NASA/ESA and STScI: p. 37tl
Robert H. McNaught: pp. 4–5, pp. 38–39
Getty Images: p. 18cr; p. 20cl
NASA/STSCi: p. 30bl
Kevin Cooper: p. 43cl

Contents

What's in the Sky? 6–7
Satellites and Space Travel 8–9
Constellations and Galaxies 10–11
The Milky Way 12–13
The Solar System 14–15
The Sun 16–17
Mercury 18–19
Venus 20–21
Earth 22–23
The Moon 24–25
Mars 26–27
Jupiter 28–29
Saturn 30–31
Uranus 32–33
Neptune 34–35
The Dwarf Planets 36–37
Comets, Meteors and Asteroids 38–39
Nebulae and Supernovae 40–41
Where, When and What to Use to See the Sky 42–43
Activities 44–45
Want to Know More? 46
Glossary 47
Index 48

What's in the Sky?

How many times have you seen the Sun rise or set? Every morning this giant star rises in the east and at dusk it sets in the west. Even though it looks like the Sun moves across the sky, it is we who are moving as Earth rotates on its **axis** and revolves around the Sun. At nighttime the sky lights up with shining stars and often we can see the Moon in its various phases.

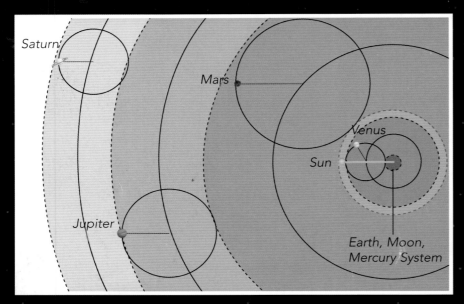

Rotating Earth

When you gaze up at the night sky, it looks like a huge black dome revolving around us, with the Moon and stars fixed to it. Earth rotates too slowly for us to feel the movement, but around it goes, making a complete turn once every 24 hours. You can track this motion by finding a bright star low in the east soon after night falls and tracing its path as it appears to move westward across the sky.

This diagram shows how our early ancestors believed that the Sun and the planets orbited around Earth. We now know this to be untrue.

Wandering Stars

It is easy to see why our ancestors thought that our planet was at the centre of all things. They believed that the Sun, the Moon and the stars all revolved around us. But even so, keen-eyed early observers noticed that some of these 'stars' moved in ways not easily explained. They called these mysterious movers 'planets', a name which means 'wanderers'.

This image shows how many hundreds of stars can be seen from dark skies in country locations.

Famous Stargazers

Many years ago, an **astronomer** called Nicholas Copernicus began to wonder about our place among the stars. He decided that Earth and the other planets revolved around the Sun. Then, about a hundred years later in 1609, Galileo Galilei (pictured) used one of the first telescopes ever made. This was the first time the Moon and some of the planets had been studied in such detail. Galileo's careful observations proved that Copernicus had been right all along.

A Star or a Planet?

Stars: Creating Light

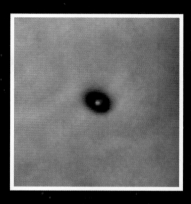

You don't need a telescope (or a bang on the head!) to see stars. On a clear night hundreds of twinkling lights are easily seen in the sky—nearly all of these are stars. Stars twinkle because they are so far away from Earth that the movement of air in the sky makes the light they give off seem to shimmer. Without a telescope planets can look a lot like stars, but they don't usually twinkle because they are much closer to Earth. The central red glow in this image is a young, newly formed star.

Planets: Reflecting Light

Planets don't produce their own light like stars do. They shine because they are reflecting sunlight from their surface, much like what would happen if you shone a torch onto a ping-pong ball in a dark room. The planets that are easiest to spot in the night sky are Venus, Mars, Jupiter and Saturn.

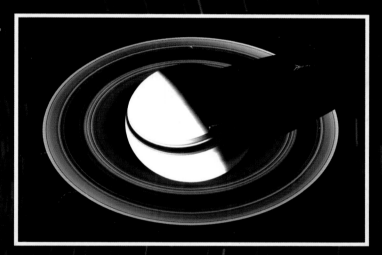

This image shows how planets (like Saturn seen from the Cassini spacecaft) shine by reflected sunlight and a shadow of the planet is cast across its rings.

As Earth rotates, the stars move across the sky during the night, just as the Sun does during the day. This photograph, taken over time, shows this movement, known as star trails.

The 'Hubble'

Satellites such as the Hubble Space Telescope are looking deeply into space. The pictures taken from the Hubble Space Telescope are much clearer than those taken from Earth and help scientists learn more about the **universe**. This image of the Cat's Eye **nebula** and its dying central star was taken by the Hubble Space Telescope.

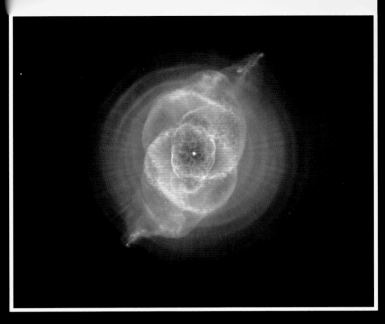

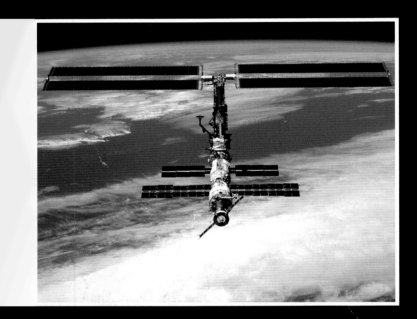

This image shows the International Space Station early in its construction. Astronauts live and work here, and communicate with Earth by radio and television transmission.

From Space to Your TV!

One type of artificial satellite is used for communication and sends, or transmits, television pictures, radio signals and telephone calls from one country to another. Other satellites are used by the military to find out what is happening in different countries around the world. You have probably seen the detailed photos that are taken by satellites in movies or on television.

Our Very Own Spaceship

The International Space Station is a huge spacecraft that orbits Earth. It has astronauts living and working on board. The International Space Station can be seen very clearly on some nights as it passes across the sky.

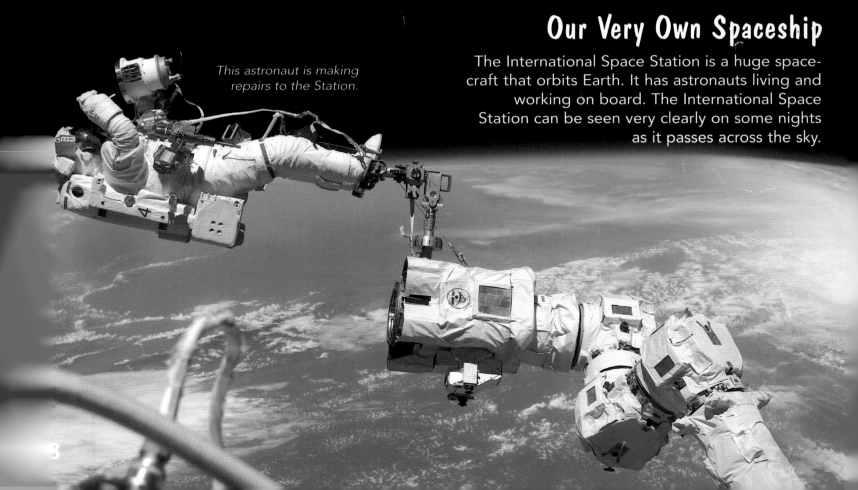

This astronaut is making repairs to the Station.

Satellites and Space Travel

If you gaze at the night sky for long enough you may see small points of light moving slowly across the sky. These are artificial satellites and there are up to 1000 of them orbiting Earth. A satellite is something that travels around a planet, like a moon. An artificial satellite is a machine built by humans and sent into space on rockets. These satellites carry out experiments and allow us to communicate with each other. We can see them because they reflect the Sun's light, just like the Moon and the planets do. As well as satellites, astronauts travel into space to collect information about our **universe**.

Dangerous Business

There are two main types of vehicles used for space exploration—rockets and shuttles. Rockets are mostly used to launch satellites and probes into space and are generally not reusable. Space shuttles are capable of taking astronauts into space for a period of time and then bringing them safely back to Earth. Despite the huge advances we have made in space travel it is still a dangerous business—in 1986 all seven crew members of the space shuttle *Challenger* were killed when the rocket carrying the shuttle exploded. And while we have made huge strides in space exploration we have still only explored a tiny part of the **universe**.

This rocket is blasting into space.

May the Force Be With You

A major challenge to space travel was achieving the powerful force needed to break free from Earth's gravity. We needed the necessary technology to produce powerful rockets that were able to travel beyond our atmosphere and into space.

Constellations and Galaxies

Like a giant spaceship, Earth orbits the Sun each year. As this occurs our view of the stars changes. Our early ancestors looked at groups of stars and imagined they could see patterns, called constellations. Most constellations are named after characters and creatures from ancient myths and legends, such as Orion the Hunter and Taurus the Bull. Galaxies are huge islands of gas, dust and stars floating together in the dark cold universe. Earth is in the galaxy known as the Milky Way.

It's All About Perspective

From where we stand it looks as though the stars that form the constellations are close to each other, but they are actually great distances apart. If you could see the constellations from some distant point in the galaxy you wouldn't recognise them—they would look very different to how we see them from Earth.

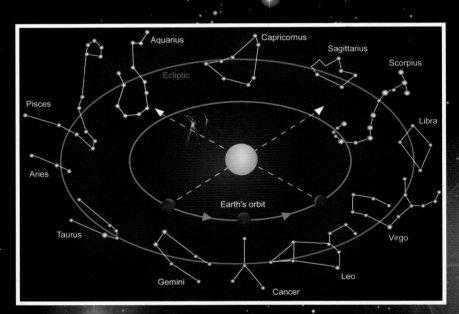

This diagram shows the patterns of the 12 common constellations.

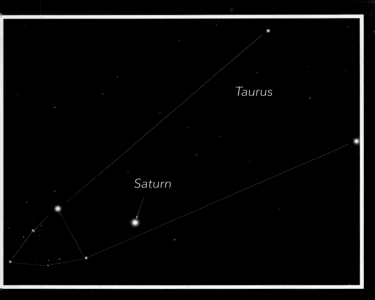

In this photo of the constellation of Taurus, can you see a planet? It looks a lot like another star, but is in fact the ringed planet Saturn.

The constellation of Orion is one of the most easily recognised star patterns in the night sky. The Orion **nebula** can be found near the middle of the Orion constellation.

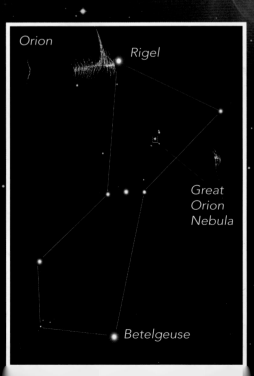

Are We There Yet?

The distance between galaxies is measured in light years, not kilometres. A light year is the distance light travels through space in one Earth year. Light travels at a speed of nearly 300 000 kilometres per second: the speed of light. The distance travelled in a light year is 300 000 (kilometres) x 60 (seconds) x 60 (minutes) x 24 (hours) x 365 (days) = 9 460 800 million kilometres—that's a long way!

While certain stars seen here are part of the colourful nebula, some are much closer or further away than it.

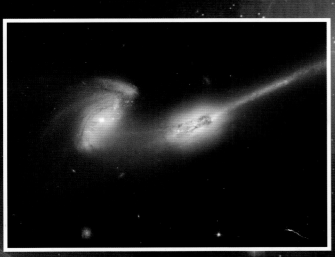

All Shapes and Sizes

Galaxies form many different shapes and sizes, from flattened discs to spheres or ovals. Some have irregular shapes, which can be caused when two galaxies collide, as shown in this image. There are even galaxies shaped like the letter 'S' which are called S-spirals. The most striking galaxies in the universe look like huge whirlpools. The outer arms of spiral galaxies contain young stars; stars nearer the centre are typically much older.

Clusters of Galaxies

In the 1920s, a man named Edwin Hubble discovered that the universe is made up of billions of galaxies. They are found in large regions called galaxy clusters, that can contain dozens or even thousands of galaxies. Each cluster is held together like a family by the forces of gravity, in the same way that the planets are held in their orbits around the Sun. Distances between the galaxies in a cluster are big, but the distances between galaxy clusters are unimaginably huge.

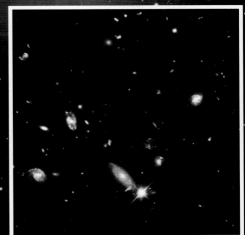

The bright lights in this photo look like stars in the night sky. But this view, taken by the Hubble Space Telescope, is of hundreds of galaxies in one very small patch of the night sky.

This photograph, taken by the Hubble Space Telescope, shows a spiral galaxy much like the Milky Way. If we could travel far off into space this is how our home galaxy might look.

Galaxies in Your Own Backyard

All the stars you can see when looking at the night sky are part of the Milky Way. The Local Group is home to the famous Andromeda galaxy and Magellanic Clouds. Less easy to see are the smaller galaxies, called dwarf galaxies, which make up the rest of the Local Group.

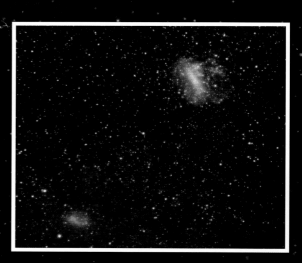

In this picture we see two small galaxies called the Large and Small Magellanic Clouds which are in orbit around our own galaxy, The Milky Way. You can see them to the south easily from dark country skies.

Galaxy Gazing

How far can you see into space? Maybe further than you think. The furthest object we can see without a telescope is the galaxy of Andromeda. All we can see is a fuzzy patch of light in the sky but that is not surprising since this giant mass of stars is 2.5 million light years away. If you know where to look, and you have a telescope, you can also gaze upon other galaxies beyond our Milky Way. These more distant galaxies belong to other clusters and you can only see them through a telescope.

The Milky Way

If you look up at the stars on a clear, dark night you may notice a faint fuzzy band of light arching across the sky. This is called the Milky Way—the galaxy in which we live. The Milky Way is only one of billions of galaxies spread across the **universe**.

A Lot of Empty Space!

The Milky Way is actually a medium-sized spiral galaxy that belongs to a small cluster called the Local Group. It contains around 100 000 million stars, including our Sun, but most of the Milky Way is empty space. If a spaceship could travel at the speed of light it would take 100 000 years to travel from one end of the Milky Way to the other! The cold gas and dust clouds it would encounter on the way are the building blocks of stars and planets.

This long exposure photograph reveals many thousands of stars but they are only a part of the Milky Way, as seen from Earth.

A Long Trip Around

Our Solar System travels around the centre of the Milky Way, but it is not a quick trip—it takes about 220 million years for the Solar System to completely orbit the centre.

of the Sun

The word 'solar' means 'of the Sun', which is why we call our planets and moons the 'Solar System'. The Sun's massive size means that it has a stronger gravitational pull than any other object in our Solar System. It is this strong gravitational pull that keeps the planets in **orbit**. Though the planets all revolve around the Sun in the same direction they travel at different speeds, with those closest to the Sun travelling the fastest.

Pluto

Neptune

Uranus

Not the Only One

Our Sun is not the only star that has planets. Since 1995, astronomers have discovered more than 120 planets orbiting other stars and have even detected an atmosphere around one of them.

Rocks and Gas

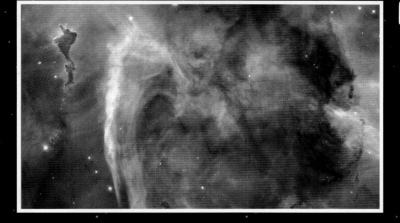

The Sun and planets were created from a huge **nebula** (like the one pictured here) of spinning gas and dust that eventually flattened into a disc over time. The inner planets—Mercury, Venus, Earth and Mars—are all rocky worlds. The outer planets—Jupiter, Saturn, Uranus and Neptune—are giant worlds of swirling gas. Pluto, and the other dwarf worlds beyond, are mostly ice.

*The **Voyager 2** spacecraft was launched in 1977 and is still making its journey through the Solar System.*

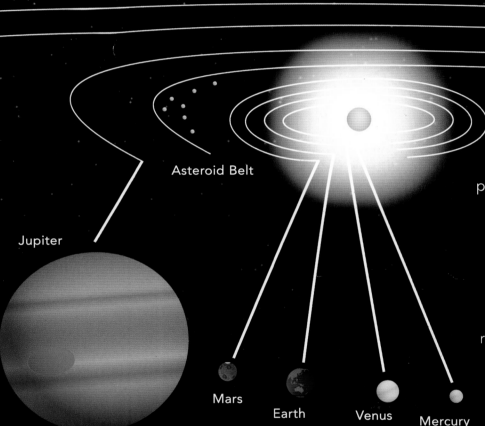

Meet the Family

The planets orbit in an almost flat plane around the Sun and in the same direction as the Sun rotates. If you were to take a journey from the Sun out towards the edge of the Solar System, you would encounter eight major planets. The first is barren Mercury; then cloud-covered Venus; our beautiful blue world, Earth; the red planet, Mars; massive Jupiter; the ringed world of Saturn; mysterious Uranus; blue Neptune; and beyond, the dwarf planets including Pluto.

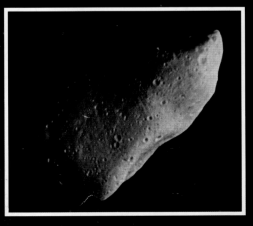

Asteroids, Ice and Comets

On your journey through the Solar System, between Mars and Jupiter, you would encounter a rocky belt of **asteroids** (pictured). Travelling beyond tiny Pluto you would find a band of small icy objects, some of which occasionally visit the inner Solar System as comets.

The Sun

The Sun is the brightest object in the sky. Like other stars, it is an enormous nuclear powerhouse with temperatures at its core reaching 15 million degrees Celsius! Yet the Sun is really a very average type of yellow star, roughly halfway through its life. Compared with the planets in our Solar System, the Sun is huge, with a **diameter** 109 times greater than Earth's and enough energy to keep burning for another five billion years.

Solar Flares

Did you know that huge explosions on the Sun can cause blackouts on Earth? Known as solar, or Sun flares, these explosions can release as much energy as ten million hydrogen bombs! The flares can leap tens of thousands of kilometres into space and the energy they release has been known to cause power blackouts on Earth. Special satellites monitor the Sun to give us an early warning that this might be about to occur.

An Eclipse of the Sun

Can day turn to night in a matter of minutes? Yes it can—when we experience a solar eclipse. An eclipse takes place when our Moon passes between us and the Sun. The Moon blocks some of the Sun's light from reaching Earth, casting a shadow that can temporarily turn day to night. A solar eclipse is an exciting event but never look at one directly. The intense light at the edges of the eclipse can cause serious and permanent damage to your eyes.

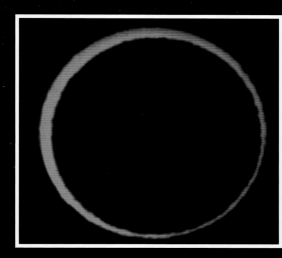

The Essentials: Light and Heat

The Sun is the nearest star to Earth and without it no life would exist on our planet—there would be no light and no heat. The light from the Sun takes more than eight minutes to reach Earth. That doesn't sound like long, but remember, light travels at nearly 300 000 kilometres a second!

Hydrogen Gas into Helium

Our Sun is constantly ejecting huge amounts of energy in violent and spectacular eruptions. Like other stars, it is made up of hot hydrogen gas. It expels around four million tonnes of material each second as it converts the hydrogen into another gas called helium.

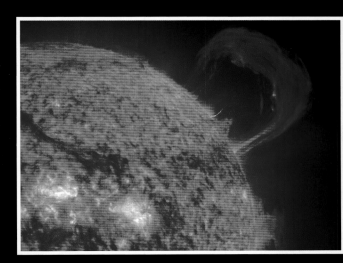

Fluid Heat

Unlike Earth, the Sun has no solid surface. The surface we can see is called the photosphere, where the temperature is around 6000 degrees Celsius. The heat we receive on Earth from the Sun has a huge effect on weather patterns around the world.

This image shows the changing pattern of heat in the Sun's atmosphere.

Mercury

Closest to the searing heat of the Sun is the small, rocky world of Mercury. Tiny Mercury is only 4880 kilometres in diameter and **orbits** the Sun at an average distance of 58 million kilometres once every 88 Earth days. Mercury looks a lot like our Moon, as it is covered with **craters** from many large meteorite impacts. Most of these collisions took place millions of years ago.

Seeing Mercury

Because it is so close to the Sun, Mercury is very difficult to see from Earth. It can sometimes be seen low in the western sky after sunset, or in the eastern sky before sunrise. If you could visit Mercury, the Sun would appear 6.3 times brighter than on Earth and just over 2.5 times as big.

Winged Messenger

Like all of the planets in our Solar System, Mercury was named after an ancient god. In Roman mythology, Mercury was the son of Jupiter and the swift messenger of the gods, recognised by his winged helmet.

A comparative view of Earth and Mercury.

A Place of Extremes

Mercury is a dead, barren world, with a large core of iron and rock. Unlike most of the planets in our Solar System, Mercury has no moon. Temperatures on this planet have been calculated to range from a blistering 427 degrees Celsius during the early afternoon to a chilly minus 183 degrees Celsius during the night—this gives Mercury the greatest temperature extremes of all the planets. Like our Moon the virtually non-existent atmosphere of Mercury means that the heat of its daytime side is lost immediately in the shadows of the night side. Surprisingly, despite being so close to the Sun, Mercury is not the hottest planet in our Solar System—neighbouring Venus holds this record.

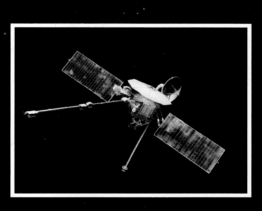

NASA Plans

Mercury has so far only been visited by one spacecraft, or probe, called *Mariner 10* (pictured here), which was launched in 1973 to commence mapping the planet's surface. But humans haven't finished with Mercury yet: in 2009 NASA will place a new probe into orbit called *Messenger*. *Messenger* will hopefully produce a complete global map of Mercury's surface and try to determine if there is any water-based ice deep in the planet's extremely cold polar craters.

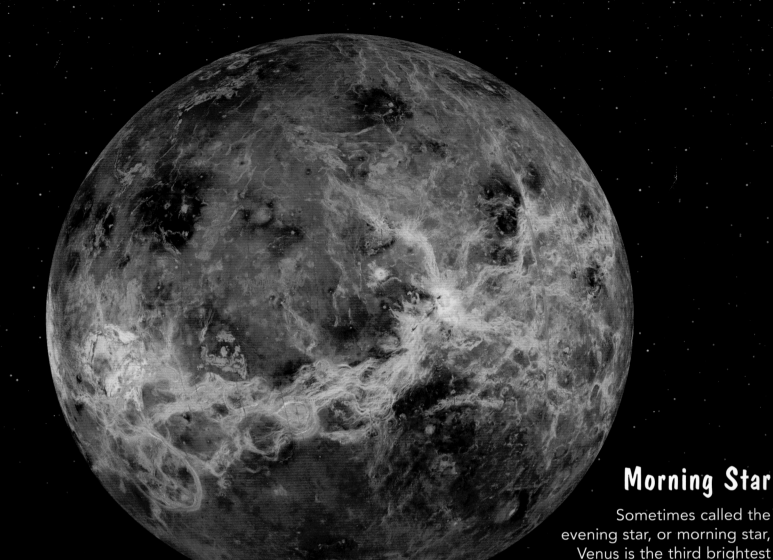

Morning Star

Sometimes called the evening star, or morning star, Venus is the third brightest object that can be seen from Earth, after the Sun and Moon. Venus is so visible because it **orbits** the Sun closer to Earth than any other planet, and its clouds are highly reflective. This is why it is easily spotted before sunrise and after sunset.

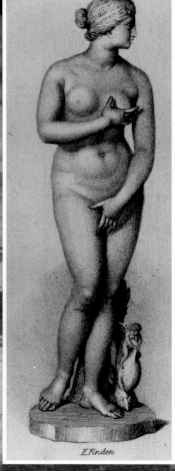

Planet of Love

Venus was the Roman goddess of love. The planet Venus is the only major planet in our Solar System to be named after a female figure.

Dry As A Bone

The surface of Venus is made up of large hot plains of dried lava and mountain ranges. Some of these mountainous features are dormant volcanoes. But unlike our watery planet, Venus has no rivers, lakes or oceans. Like Mercury, Venus has no moon of its own. **Orbiting** 108 million kilometres away from the Sun spaceprobes sent to Venus, such as the *Magellan* spacecraft, have provided us with spectacular images of the planet's clouds and its surface.

Venus

Similar in size to Earth, the planet Venus is completely covered by thick poisonous clouds made up of carbon dioxide gas and droplets of **sulfuric acid**. A rainstorm on Venus would do much more than make you wet—it would dissolve your clothes and burn your skin. Venus mysteriously rotates on its **axis** in the opposite direction to all the other planets. This may be due to a collision with another rocky world when the Solar System was forming.

A comparative view of Earth and Venus.

Long, Long Days

Can a day be longer than a year? It can on cloudy Venus. This planet turns so slowly that one full day on Venus is 243 Earth days. Venus' daily rotation actually takes longer than one Venus year, which is only 225 Earth days.

Trapped Heat

Because the planet is covered by layers of thick cloud it is difficult for the heat rising from the surface to escape. This makes Venus the hottest planet in the Solar System. Space probes sent to Venus found the air pressure on its dry, volcanic surface to be a crushing 90 times heavier than on Earth, with blistering temperatures reaching 480 degrees Celsius—hot enough to melt lead! With pressures and temperatures this high, humans aren't ever likely to pay Venus a visit.

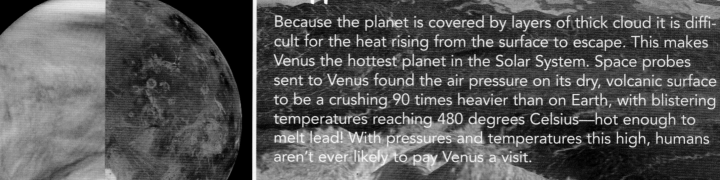

This image shows the difference between Venus covered by cloud (left) and Venus without cloud cover (right).

Earth

We've still got a long way to go on our journey through the Solar System, but let's stop for a moment's rest at home. Earth is the beautiful blue world on which we live, and the third planet **orbiting** the Sun, which is about 150 million kilometres away. Unlike most of the other planets, Earth has only one moon.

A Year Around the Sun

Our planet races around the Sun at an average speed of just over 107 000 kilometres an hour. Even so, it takes 365 days (or one full year) for Earth to make its way once around the Sun. While it is **orbiting** the Sun, Earth is also rotating on a slightly tilted **axis** once every 24 hours, or one Earth day. Because of this tilt, the amount of sunlight reaching parts of Earth's surface varies, giving us the seasons we experience. This is why when it is summer in the northern hemisphere it is winter in the southern hemisphere, and vice versa.

This diagram shows Earth's orbit over one year. At different points along its orbit around the Sun, each hemisphere receives more or less direct sunlight, giving us the seasons we experience.

All On Our Own

You've probably seen movies or read books about monsters from space and invaders from Mars. But, to the best of our knowledge, Earth is the only planet where living things survive. Earth is an ideal home for plants and animals because it provides two of life's essentials—oxygen to breathe and water to drink. And our planet is just the right distance from the Sun to not be too hot, like Venus; or too cold, like Mars.

Blue Ball

When seen from space, Earth looks like a giant blue ball because of the water on its surface. Earth is unique in the Solar System because of this liquid water, which covers almost 70 per cent of the planet's surface. The other planets have either lost whatever water they once had, or are so cold that water only exists as ice.

Protecting Atmosphere

Earth's **atmosphere** is a thick layer of gases which let in the Sun's light and heat. It is a big part of how and why so much life can exist on Earth. Our atmosphere protects us from many of the harsh rays of the Sun and contains the oxygen we need to breathe. The further you travel away from the surface of Earth, the less oxygen there is. Once you leave Earth's atmosphere there is no oxygen left and what we call 'space' begins.

Lunarscape

The Moon's surface is littered with rocks and **craters**. The bright areas we can see from Earth are heavily cratered regions of high ground, while the darker areas are dry, frozen plains of ancient **lava**.

Faces of the Moon

The Moon spins very slowly on its **axis**, taking 27.3 days to complete one rotation, which is the same time it takes it to orbit Earth once. This is why we always see the same face of the Moon. As it revolves around Earth, the Moon's appearance changes nightly and we see more or less of its sunlit side. Some nights we see it as a crescent, sometimes half-lit, and sometimes as the full Moon.

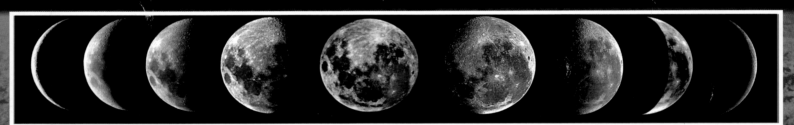

Ebbs and Flows

Although the Moon is smaller than Earth, it is large enough and close enough to have a gravitational effect on our oceans, giving us high and low tides each day.

This image shows a comparative view of our big planet and much smaller Moon.

The Moon

The Moon is our nearest neighbour and the only place in the universe (apart from Earth!) where humans have walked. The Moon is about 384 400 kilometres away from us and there is less **gravity** there than on Earth. Because of this you could jump about six times higher on the Moon because you would only weigh one-sixth your normal weight. On the Moon, you'd be able to beat Olympic high jumpers—as long as they were still on Earth!

Moon Walking

You probably weren't even born when humans took their first steps on the Moon but it was one of the most watched and anticipated events in human history. Over 500 million people around the world looked on in 1969 as astronaut Neil Armstrong became the first human to step onto something that people had been gazing at for thousands of years, and heard him say, 'That's one small step for a man, one giant leap for mankind'. Since then scientists have learnt a lot about it.

This photo shows astronaut Eugene Cernan travelling over the surface of the Moon in a lunar roving vehicle.

Six Trips to the Moon

Since the first mission to the Moon there have been five more, with the last taking place in 1972. Twelve astronauts have now walked on its surface, spending six days travelling through space to get there and back.

Mars

Named after the Roman god of war, Mars (the fourth planet from the Sun) is sometimes called the 'red planet' because of its reddish appearance. This is caused by a rusting effect in its surface soil and rocks, and led to the ancient Egyptians naming it the 'Red One'. Mars **orbits** about 228 million kilometres from the Sun.

A comparative view of Earth and Mars.

Looking like something from a science fiction movie, this rover explorer roams the deserts of Mars.

Roving Robots

Mars is the best-explored planet in our Solar System. Two roving robots, called *Spirit* and *Opportunity*, landed on Mars in 2004. They were sent to Mars to collect information that may tell scientists whether some kind of life once existed on this cold and unfriendly planet.

Giant Volcanoes

Did you know that there are volcanoes on other planets? There are some huge dormant volcanoes on the surface of Mars including the giant Olympus Mons (pictured) which, at three times taller than Earth's Mount Everest, is the tallest known object in the Solar System.

Martian Seasons

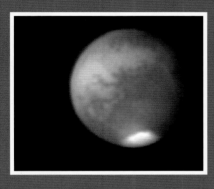

One day on Mars is 37 minutes longer than an Earth day. Mars has seasons during its year just like Earth because its **axis**, too, is slightly tilted. However, seasons on Mars (known as martian seasons) last much longer than on Earth because Mars takes nearly twice as long to **orbit** the Sun—687 days. Mars has ice caps at its north and south poles, and a very thin atmosphere, but the similarities to Earth stop there.

Dust and Cyclones

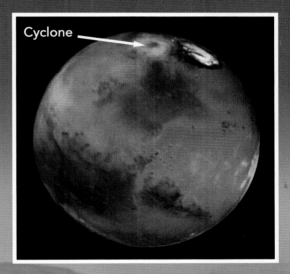

Cyclone

The atmosphere on Mars is not safe for humans to breathe and if astronauts did roam on its surface they would need to wear special spacesuits. It would take them six months to travel from Earth, and when they arrived the astronauts would find themselves in an alien climate where huge dust storms and cyclones sometimes blanket the entire planet for days or weeks. The average temperature on Mars is a frosty minus 63 degrees Celsius and at night this can drop to over minus 120 degrees.

Tiny Moons

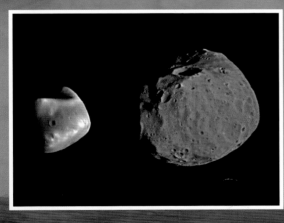

Mars has two tiny moons called Phobos and Deimos which look like potato-shaped boulders. Because of their small size it is believed that they may once have been asteroids that were pulled into orbit by Mars' gravity.

Many Moons

Jupiter has 62 known moons and more will possibly be discovered, making the planet a virtual Solar System in miniature. While most of these **orbiting** moons are very small, the four largest—Io, Europa, Ganymede and Callisto—can be seen with binoculars from Earth.

The small black dot you can see on this image of Jupiter is the shadow of the moon Europa as it passes in front of the giant planet.

Io

Europa

Ganymede

Red and Yellow Swirls

The bright red and yellow clouds of Jupiter's **atmosphere** make it a very colourful planet. Easily spotted through a telescope is the Great Red Spot—an oval-shaped cyclone or storm that has been raging for over 300 years. This storm could swallow up two or three Earths.

Callisto

28

Jupiter

Moving away from the Sun, far beyond Mars and the Asteroid Belt, we come to the realm of the gas giants. Placed fifth from the Sun, enormous Jupiter **orbits** at an average distance of 778 million kilometres. The largest of all the planets in the Solar System, it takes Jupiter almost twelve Earth years to make one orbit of the Sun.

Wider than Eleven Earths!

This huge planet spans nearly 143 000 kilometres in **diameter** at its equator, the equivalent of eleven Earths stacked side by side. Except for the Sun, gigantic Jupiter accounts for more than two-thirds of the material that formed all the planets in the Solar System. It would take more than 1300 Earths to fill its volume and all the other planets in our Solar System could fit inside Jupiter. Not surprisingly, given its immense size, the planet is named after the main god in Roman mythology.

This comparative view of Earth and Jupiter shows just how big this gas giant is.

Fast Days and Stormy Nights

Huge bands of cloud and circular storms can be seen moving swiftly through Jupiter's atmosphere. That's not the only fast thing about Jupiter—it spins swiftly on its axis and one day is less than ten hours long.

Moons Bigger than Planets

Jupiter's largest moon, Ganymede (pictured opposite), is the largest of all the moons in the Solar System and is even larger than the planet Mercury. But because it orbits Jupiter and not the Sun it cannot be classed as a planet. Io (also pictured opposite) is the most volcanic body in the Solar System and is littered with large vents, spewing molten sulfur over its surface.

Saturn

Saturn is the most beautiful planet to observe through a telescope because of its striking rings and rich golden yellow colour. The sixth planet from the Sun, Saturn is the second largest in the Solar System.

Spinning Gas

Saturn is about 1425 million kilometres away and takes 30 years to make one orbit around the Sun. It is 752 times the volume of Earth and, like Jupiter, is a huge, rapidly spinning ball of mostly hydrogen gas. One day on this planet lasts ten hours and fourteen minutes.

A comparative view of our planet and Saturn gives an idea of its immense size.

That's a Long Way Away!

If you could drive to Saturn in a car, travelling non-stop in a straight line and doing 100 kilometres an hour, it would take you about 1365 years to reach the planet.

This image shows the beautiful, magical rings of Saturn as seen from the Hubble Space Telescope.

Mysterious Moon

Saturn has 59 moons, of which Titan is the largest. Titan is the second largest moon in the Solar System after Jupiter's Ganymede. It is covered by a thick orange atmosphere which until recently had hidden views of its surface from the prying eyes of spacecraft. The *Cassini* mission of 2004 used special cameras to unveil this hazy moon. *Cassini* also released a smaller probe called *Huygens*, which landed on Titan in January 2005.

The special cameras aboard Cassini *peered through the thick clouds that surround Saturn's largest moon Titan, in January 2005, to take this photograph.*

Saturday Named from Saturn

Saturn was named after the Roman god of agriculture and father of Jupiter. In ancient times Saturn was the furthest planet from us that was known to human observers. The first day of our weekend takes its name from Saturn.

Wind and Ice

The winds on Saturn can travel at speeds of up to 1440 kilometres an hour. Through a large telescope you can see faint bands of gaseous clouds across its globe. The famous 'rings of Saturn' are made up of millions of small icy and rocky bodies. The rings cover a distance over 275 000 kilometres from edge to edge, but are only about one kilometre thick. The icy chunks of rock reflect the Sun's light, giving the rings their dazzling appearance.

Uranus

The seventh planet from the Sun—a distance of 2870 million kilometres—is Uranus. Unlike the other planets, Uranus is tipped on its side, perhaps as a result of some catastrophic event early in its history. Because of this tilt Uranus rolls like a wheel as it **orbits** the Sun.

Green Disc in the Sky

Uranus takes 84 Earth years to make one **orbit** of the Sun, but one day on Uranus lasts only seventeen hours. The planet is so far away from Earth that it appears as little more than a tiny greenish disc through a small telescope and it took *Voyager 2* over eight years to reach it. Despite being such a long way away Uranus can be seen without a telescope on clear, moonless nights. But you will need a **star chart** to know exactly where to look.

This image of Uranus and a handful of its moons was taken with a small telescope from a backyard on Earth.

Hydrogen World

Uranus is a gas giant like Jupiter and Saturn, made up mostly of hydrogen gas. But Jupiter and Saturn are much bigger worlds in comparison. Uranus has a **diameter** of 51 118 kilometres and is 64 times the volume of Earth (see below). The upper cloud on Uranus is hazy and pictures from the *Voyager* spacecraft revealed a bland-looking, greenish blue world.

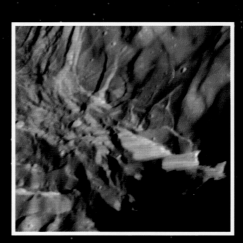

This image shows the highly disturbed cliffs and canyons on the surface of Uranus' moon Miranda seen from the Voyager 2 spacecraft.

Cold Moons

Uranus has 27 moons. The five largest are Titania, Oberon, Umbriel, Ariel and Miranda. Tiny Miranda, only 480 kilometres in diameter, has huge ice cliffs that tower high above deep canyons. At this distance from the Sun the average temperature on Miranda is a very cold minus 205 degrees Celsius.

Seasons that Last for Years

Uranus has up to thirteen thin, dark rings. Because it is tilted on its side, half of Uranus receives the Sun's light while the other half remains in darkness. This makes seasons on Uranus last for years instead of the months we are used to on Earth.

This Hubble Space Telescope image of Uranus shows its delicate rings.

Roman Connections

Uranus is the only planet in our Solar System to be named after a Greek instead of a Roman god. In mythology Uranus was the Greek god of the sky, a fitting character to name this gaseous planet after.

Uranus was discovered in March 1781 by the English astronomer Sir William Herschel.

Reversing Moons

Neptune (pictured above) has thirteen known moons, of which Triton is the largest. This strange moon **orbits** the planet in the opposite direction to the other moons and may have been an independent world that was captured by Neptune's strong gravity long ago. *Voyager 2* revealed that this moon is 2706 kilometres in **diameter** and has active **geysers** on its icy surface that eject huge plumes of dark material high into the Triton sky.

A comparative view of Earth and Neptune, the blue planet fittingly named after the Roman god of the sea.

Small Sun

If you visited Neptune you would notice that the Sun is very small compared with its size as seen from Earth. The light given off by the Sun is also much dimmer—about a thousand times less than we see it from Earth.

Earth, Neptune appears as little more than a bluish-coloured dot when viewed through a small telescope. This picture of Neptune and its largest moon, Triton, was taken with a large telescope and shows how smaller Triton has moved around Neptune.

Rings and Clouds

It took twelve years for the *Voyager 2* spacecraft (pictured here) to reach Neptune and in 1989 it returned the first close-up pictures ever seen. The spacecraft revealed that Neptune has rings but they are faint. It also photographed a large and mysterious dark spot, which has since disappeared, and wispy white clouds in the planet's **atmosphere**.

Bad Reception?

Neptune is so far from Earth that it took more than four hours for transmissions from *Voyager 2* to reach us, even though they were travelling at the speed of light! The transmissions were also very weak and it took every bit of radio dish receiving power here on Earth to pick them up.

This image shows a satellite dish which picks up transmissions from outer space.

Windy Planet

Neptune is the windiest planet in our Solar System and **Voyager 2** recorded wind speeds over 1600 kilometres per hour. In the most destructive cyclones on Earth, wind speeds rarely reach more than 280 kilometres per hour.

The Dwarf Planets

Now we really are a long, long way from home. Until recently, Pluto was always considered to be the ninth planet. In 2006 the definition of 'planet' was changed and poor old Pluto is now considered only to be a dwarf planet.

Bound Together

Discovered by the American astronomer Clyde Tombaugh in 1930, Pluto is one of the coldest objects in our Solar System—an icy world that **orbits** at about 5905 million kilometres from the Sun. The moon of Pluto, called Charon, is little more than half the size of Pluto itself. Bound together by gravity, the pair takes 248 years to orbit the Sun. This means that one year on Pluto is equal to 248 Earth years.

The Kuiper Belt

The Kuiper Belt is a vast plane of rock–ice objects called **planetesimals**, much like an outer asteroid belt. The Kuiper Belt extends far beyond the **orbit** of Pluto. Many comets originate from the Kuiper Belt.

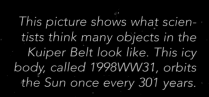

This picture shows what scientists think many objects in the Kuiper Belt look like. This icy body, called 1998WW31, orbits the Sun once every 301 years.

Spotting Pluto

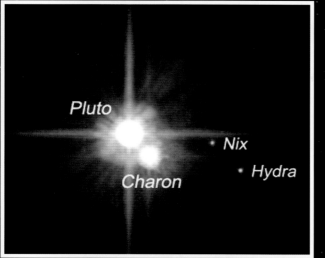

Pluto and its moons Charon, Nix and Hydra, are extremely cold worlds with icy surfaces that may have rocky cores. In 1988 astronomers discovered that Pluto has a thin **atmosphere** of hydrogen gas. It is probably no surprise that Pluto can be difficult to find, let alone see. It can be spotted, but this requires a detailed **star chart** and a powerful telescope to locate it among the stars.

Odd Orbits

Pluto's odd orbits means that it sometimes passes inside the orbit of neighbouring Neptune. At its most distant Pluto can be 7524 million kilometres from the Sun but in 1989 it came as close to the Sun as it can—just 4437 million kilometres! This image shows Pluto's passage over 8 days.

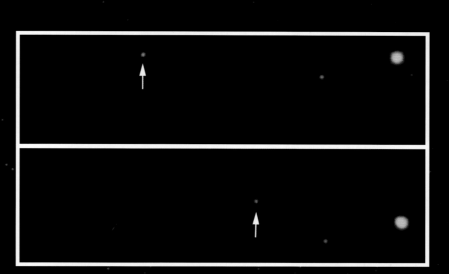

A comparative view of our Earth and small, cold Pluto.

Underworld Planet

Pluto is named after the Roman king of the Underworld, the shadowy land where the souls of the dead lived. One of its three moons, Charon, is named for the ferryman who rowed the newly dead across the Underworld's River Styx. Sometimes people were buried with coins under their tongue to pay the ferryman for their last journey.

The Asteroid Belt

Most **asteroids** are found in a region between Mars and Jupiter called the Asteroid Belt. They range from under a kilometre in size to the largest, Ceres, which is 940 kilometres in **diameter**.

Flying Comets

While comets move quickly in space, we don't see them streak across the sky because they are a long way from Earth. A comet's brightness as seen from Earth depends on its size and distance.

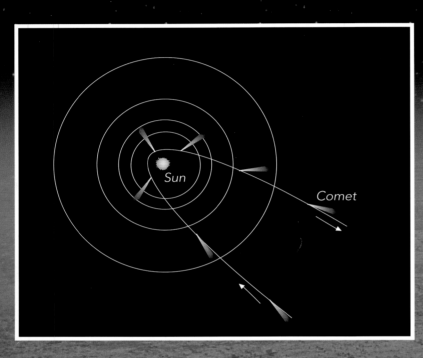

As this diagram shows, a comet's tail (shown here in blue) always points away from the Sun.

Showers of Light

Sometimes Earth passes through a band of tiny dust particles left behind by a comet. From Earth, this appears as a spectacular meteor shower with hundreds of light streaks shooting through parts of the sky.

Halley's Comet

Comets visit the inner Solar System regularly, like the famous Halley's Comet which returns every 76 years—astronomers call these short-period comets. Icy comets from the Oort Cloud can zoom in from almost any direction and return every thousand years or so—these are called long-period comets.

Comets, Meteors and Asteroids

This image of the comet Levy streaking through the sky was photographed from Earth.

When thinking about space the beautiful planets take centre stage—but we can't forget about comets, meteors and asteroids! Comets originate from a region called Kuiper Belt and another more distant place called the Oort Cloud (a spherical halo of icy objects surrounding the Solar System). Sometimes they are influenced by the **gravity** of larger objects and can begin travelling inwards toward the Sun. Heating up along the way, they can form long tails and sometimes even collide with the Sun or a planet like Jupiter. Meteors, however, are small pieces of rock, metal or ice that travel through space at enormous speeds, only to burn up as they enter Earth's **atmosphere**. Asteroids are rubble fragments remaining from the formation of our Solar System.

Look Out!

If you've ever seen a shooting star, then you've seen a meteor. Most meteors burn up completely but big ones can reach the ground and have been known to hit cars and houses. When they do hit the ground and survive they are called meteorites. This **crater** in the Northern Territory was created long ago by a meteorite's impact.

The Main Players

The term asteroid means 'star-like'—from Earth asteroids appear like tiny points of light moving slowly among the stars. The four largest asteroids in our Solar System are Ceres, Pallas, Juno and Vesta. The first to be discovered was Ceres, in 1801, and more are being discovered all the time. Asteroids may be made out of rock or metals, and most of them look like massive boulders floating in space. The biggest of them look like small planets and so they are called the minor planets!

Fading Away...

Smaller stars die off in a much gentler way than supernovae. Rather than exploding they slowly cool, shrink and fade away.

Bursting with Energy

So much energy is released in a supernova event that one can easily outshine the combined brilliance of all the stars in an entire galaxy. After the explosion, a small core of the original star may remain. This is known as a neutron star or a pulsar.

Even Stars Die

Black holes can be formed when a large star dies. **Gravity** can exert so much pull on the fragments of a dead star that they are drawn together, sucking in everything nearby, including light. This creates a black hole, from which nothing can ever return. Black holes are known to exist at the centre of many galaxies, including our own Milky Way.

*The Lagoon **Nebula** in Sagittarius. In nebulae like this one new solar systems evolve within dark and cool collapsing clouds.*

Shedding its Skin

Another type of **nebula**, called a planetary nebula, is created when an ageing star sheds its outer layers into space, forming a cocoon of gas and dust around the small central star. Swarms of gas and dust form this beautiful nebula where new stars and planets are being born deep within.

Nebulae and Supernovae

Nebulae are the beautiful clouds of dust and gas that are found among the stars within galaxies. Through a telescope, nebulae look like hazy patches of light, and some can even be seen with the naked eye. When some massive stars have exhausted the fuel that has kept them burning for billions of years they start to become unstable. These stars, which are about eight times the size of the Sun, begin to collapse in on themselves, causing a catastrophic explosion that is called a supernova.

The Orion and Horsehead Nebulae

Many nebulae are brightly lit, or illuminated, by the young stars that are being formed within them, like the famous Orion nebula. Others are illuminated by reflected light from nearby stars. Some nebulae appear dark because they are in front of a brighter nebula, such as the Horsehead nebula, also in the constellation of Orion.

This image shows the swirling dark cloud of the famous Horsehead nebula.

Timing is Everything

Try to avoid setting up your telescope (like the one pictured here) on heat absorbing surfaces such as concrete or pointing it over rooftops, as the heat rising from these surfaces can seriously reduce the visual quality. Also, there are times for observing that are better than others—generally speaking just before sunrise, just after sunset and around midnight are the best.

Only the Beginning?

One of the best objects to start observing is our Moon (pictured here). Even a good pair of binoculars will reveal its mountainous highlands and low-lying dark plains, and the most basic of telescopes will bring these features into even greater detail. You can finetune your observation and technical skills on the Moon and then start seeking more difficult subjects like the planets. It may lead to a career, or a rewarding lifelong interest, but for astronomers worldwide the sky is truly a beautiful place to explore.

Some Extra Help

Finding what you are looking for amongst the mass of stars in the night sky can be difficult, even for experienced astronomers. Luckily, **star charts**, websites and computer software (some of which can be downloaded for free) can all help with your star and planet spotting. Many astronomy magazines publish monthly night sky charts or you can visit some of the websites on page 46. Having some of this background information also helps you learn about tools (like the complicated piece of equipment shown here) to use as you become more knowledgable about the universe.

Where, When and What to Use to See the Sky

Over a thousand stars and six of the planets—Venus, Jupiter, Saturn, Mars and, to a lesser extent, Mercury and Uranus—are visible to the naked eye but it is possible to see a lot more up there if you have some guides, and know just where to point your binoculars or telescope.

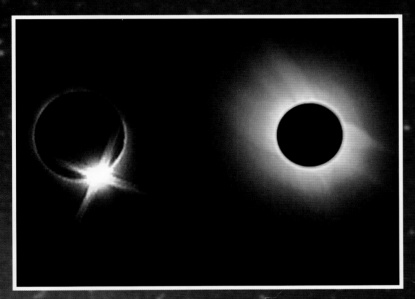

This image of an eclipse was taken with a special solar filter. You should never look directly at the Sun because it can damage your eyes.

Some Fun Homework

When it comes to purchasing your first telescope, you should start by asking yourself a few simple questions: Can I use it for taking photographs? Will I clearly see the rings of Saturn 1.5 billion kilometres away? Do I need to transport it a lot and will it fit in the family car? What accessories will I need and does the telescope depend on a power supply? Even the most basic telescopes are not cheap, so it is important that you buy one that is right for you. Work with your parents or friends to make a list of questions to ask and do some research online or at the shop. This way you'll avoid the disappointment that often comes from making a hasty decision.

Calling All Urban Astronomers!

You might think that you need to be out in the middle of the desert or high on a mountaintop to see the night sky clearly but this is not necessarily so. It is true that the less light there is around you, the clearer the sky will be, but city living needn't restrict your excursions into astronomy. You can use simple binoculars (pictured here) from your balcony or backyard to observe the night sky because most of the easiest objects to spot are bright enough or reflect enough light to still be seen from urban areas.

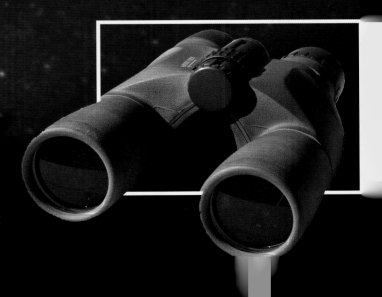

Activities

The answers to these questions are all in this book. To help you, the page numbers you should look at are given for each.

1. How often does Earth go around the Sun? (pages 6–7)
2. Do planets produce their own light? (pages 6–7)
3. What is an artificial satellite? (pages 8–9)
4. How many constellations can you see in the pictures on page 10. What are their names? Do the names seem familiar? (pages 10–11)
5. What is a galaxy? Which galaxy is Earth in? (pages 10–11)
6. How fast is the speed of light? (pages 10–11)
7. What is the Milky Way? (pages 12–13)
8. What does the word 'solar' mean? (pages 14–15)
9. How many planets are there in our Solar System? (pages 14–15)
10. Why is the Sun important to Earth? (pages 16–17)
11. How did Mercury get all of its craters? (pages 18–19)
12. Have any spacecraft been near Mercury? (pages 18–19)
13. Does Venus have any moons? (pages 20–21)

14. Is Venus a dry or wet planet? (pages 20–21)

15. Why does Earth look blue from space? (pages 22–23)

16. Why is it difficult to breathe when you leave Earth's atmosphere? (pages 22–23)

17. When did man first land on the moon? (pages 24–25)

18. How many astronauts have walked on the moon? (pages 24–25)

19. What is Mars sometimes called? (pages 26–27)

20. How long does it take Jupiter to orbit the Sun? (pages 28–29)

21. What are the names of some of the moons of Jupiter? (pages 28–29)

22. How big is Saturn? (pages 30–31)

23. What are the rings of Saturn made of? (pages 30–31)

24. Why is Uranus different from the other planets? (pages 32–33)

25. When was Neptune discovered? (pages 34–35)

26. How far away is Pluto from the Sun. Does this distance change? (pages 36–37)

27. What is one of the most famous comets? (pages 38–39)

28. What is the difference between nebulae and a supernova? (pages 40–41)

29. What are black holes? (pages 40–41)

Want to Know More?

Hopefully this book has made you more interested in the fascinating, changing world above you. The links on this page won't literally take you into outer space, but they will help you know more about the wonders above and how to go about viewing them. Some of the links have fun interactive games; others are online planetariums where, once you key in your position on Earth, you can see what is in the sky right above your head. Other links are to space agencies such as the USA's NASA, where you can find out lots more information and see some amazing outer space images. And since astronomy has the largest number of amateur followers, there are heaps of groups and organisations to help you with advice on getting started with the stars.

Websites

http://www.fourmilab.ch/yoursky/
An interactive planetarium site

http://library.thinkquest.org/3461/
The Online Planetarium Show

http://www.scitech.org.au/index.php?option=com_content&task=view&id=96&Itemid=103
The Australian Online Planetarium has a section called *The Sky Tonight* to tell you what the Moon is doing and where the planets can be spotted. Also has a section on what can be seen with binoculars or a small telescope

http://www.museum.vic.gov.au/planetarium/links1.html
Links from the Melbourne Planetarium

http://home.vicnet.net.au/~apsweb/
Australasian Planetarium Society

http://www.nasa.gov/
NASA Home Page

http://hubblesite.org/
Hubble Space Telescope page

http://space.jpl.nasa.gov/
The Solar System Simulator

http://starchild.gsfc.nasa.gov/docs/StarChild/StarChild.html
A learning centre for the young astronomer

http://www.windows.ucar.edu/
Windows to the Universe—lots of interactive stuff

http://www.kidsastronomy.com/
Another site with heaps of news and views for the younger astronomer

http://www.nasa.gov/audience/forkids/kidsclub/flash/index.html
NASA Kids' Club homepage

http://asa.astronomy.org.au/become.html
How to Become an Astronomer from the Astronomical Society of Australia—more for adults, but if you really want to become an astronomer this will help you work out where to start and what you'll need to know

www.myastroshop.com.au/astroguide.asp
For buying telescopes, binoculars and other accessories to help you view the night sky.

Glossary (what words mean)

asteroid	A planetoid (a minor planet) whose orbit lies mainly between those of Mars and Jupiter.
astronomer	A scientist who studies the universe and its mysteries.
atmosphere	The layer of gases that surrounds the body of a planet or moon.
axis	An imaginary line running through the north and south poles of a planet, about which a rotating body (such as Earth) turns.
constellation	A group of stars to which specific names have been given.
crater	A rounded, hollow dent formed in the surface of a solid body like a planet or moon, caused by the impact of a meteorite or other small solid body.
diameter	The measure of a straight line passing from side to side of a body, through its centre.
galaxy	A system of stars held together by gravity and separate from other systems by large areas of space.
geyser	A small vent or crack in the surface of a planet or moon from where liquid or gases eject upwards.
gravity	The force of attraction where two or more bodies are drawn together. For example, the force that keeps everything down on Earth; or a meteorite in space being pulled in by the greater gravitational force of a much larger body, like Earth.
lava	The hot, melted fluid rock that erupts from a volcano.
NASA	National Aeronautic and Space Administration.
nebula	A cloud of gases and dust where stars and planets are formed.
orbit	The path that a planet or satellite makes around a body, such as Earth around the Sun, or the Moon around Earth.
planet	A body revolving around the Sun, or revolving around another star. Planets are only visible by reflected light.
planetesimal	A very small body (perhaps a few millimetres to a kilometre in size) that condensed from the gases and dust of the early Solar Nebula. Moving together, they gradually come together to form the planets and satellites of the Solar System. Many of these small objects still roam aimlessly throughout the Solar System.
probe	A spacecraft that is able to explore, examine and test conditions of other bodies in space and radio back the results.
solar system	The Sun together with all the planets, satellites, asteroids and other bodies, revolving around it.
star	Stars are enormous, hot, fiery balls of gas like the Sun but so far away that they appear only as points of light in the night sky.
star chart	A map of the night sky used to locate and identify stars, planets, constellations and galaxies.
sulfuric acid	A highly corrosive, oily fluid made up of hydrogen, sulfur and oxygen.
universe	Space and everything in it including our Solar System, the stars and all the galaxies.

Index

Andromeda galaxy 12
animals 22
Ariel 33
Armstrong, Neil 25
asteroid belt 38
asteroids 15, 38–39
astronauts 8, 25
astronomers 6, 32, 36
atmosphere 9, 23
black holes 40
Callisto 28
Cassini mission 31
Cat's Eye nebula 8
Ceres 39
Cernan, Eugene 25
Challenger space
  shuttle 9
Charon 36–37
charts, star 42
comets 15, 38–39
constellations 10–11
Copernicus, Nicholas 6
craters, meteorite 39
cyclones 27, 29
days
  Jupiter 29
  Mars 27
  Saturn 30
  Uranus 32
  Venus 21
Deimos 27
dwarf planets 36–37
Earth 22–23
  atmosphere 23
  axis 6
  gravity 9
  orbit 22
  position in Solar
    System 14–15
  rotation 6, 22
  seasons 22
Europa 28
famous stargazers 6
galaxies 10–11
Galilei, Galileo 6
Ganymede 28–29
gas 15, 17, 32
geysers 34

gravitational effect of
  Moon 24
Great Red Spot 28
Halley's Comet 38
helium 17
Herschel, Sir
  William 32–33
Horsehead nebula 41
Hubble, Edwin 11
Hubble Space
  Telescope 8
Huygens probe 31
Hydra 37
hydrogen gas 17, 32
ice 15
International Space
  Station 8
Io 28–29
Juno 39
Jupiter 14–15, 28–29
Kuiper Belt 36
Lagoon nebula 40
life on planets 22
light
  speed of 11, 17
  from Sun 17
  years 11
Local Group 12–13
lunar phases 24
Magellanic Clouds 12
Mariner 10 19
Mars 14–15, 26–27
Mercury 14–15, 18–19
meteor showers 38
meteors 38–39
Milky Way 12–13
Miranda 33
Moon 24–25
  gravitational effect 24
  missions 25
  observing 42
  phases 24
  rotation 24
  walking on 25
moons
  Jupiter 28
  Mars 27
  Neptune 34–35

Pluto 36–37
Saturn 31
Uranus 33
morning star 20
nebulae 40–41
  Cat's Eye 8
  Horsehead 41
  Lagoon 40
  Orion 10, 41
  planetary 40
Neptune 14–15, 34–35
neutron stars 40
Nix 37
Oberon 33
Olympic Mons 27
Oort Cloud 38–39
Opportunity robot 26
Orion constellation 10
Orion nebula 10, 41
oxygen 22–23
Pallas 39
Phobos 27
photosphere 17
planetary nebulae 40
planetesimals 36
planets
  dwarf 36–37
  life on 22
  shining 7
  Solar System 14–15
plants 22
Pluto 14–15
probe 19, 31
pulsars 40
red planet 26
rings of Saturn 30–31
robots, roving 26
rockets 9
rocks 15
rovers 26
satellites 8–9
Saturn 10, 14–15, 30–31
seasons
  Earth 22
  Mars 27
shuttles 9
sky charts 42
solar eclipse 17

solar flares 17
Solar System 14–15
space station 8
space travel 8–9, 25
spacecraft
  Cassini 31
  Mariner 10 19
  Voyager 2 14, 32, 35
spiral galaxies 11, 12
Spirit robot 26
stars
  charts 42
  dead 40
  trails 7
  twinkling 7
  wandering 6
sulfuric acid 21
Sun 16–17
  eclipse 17
  explosions 17
  gas 17
  gravitational pull 15
  light from 17
  rising and setting 6
  size 16
  surface 17
  temperature 16
supernovae 40–41
Taurus 10
telescopes 42–43
  Hubble Space
    Telescope 8
tides 24
Titan 31
Titania 33
Tombaugh, Clyde 36
Triton 34–35
Umbriel 33
Uranus 14–15, 32–33
Venus 14–15, 20–21
Vesta 39
volcanoes 27
Voyager 2 14, 32, 35
water 22–23
wind 35